Landforms:
The Ever-changing Earth

by Emily Sohn and Adam Harter

Chief Content Consultant
Edward Rock
Associate Executive Director, National Science Teachers Association

NorwoodHouse Press
Chicago, Illinois

Norwood House Press
PO Box 316598
Chicago, IL 60631

For information regarding Norwood House Press, please visit our website at
www.norwoodhousepress.com or call 866-565-2900.

Special thanks to: Amanda Jones, Amy Karasick, Alanna Mertens, Terrence Young, Jr.

Editors: Barbara J. Foster, Diane Hinckley, Michelle Parsons
Designer: Daniel M. Greene
Production Management: Victory Productions, Inc.

Paperback ISBN: 978-1-60357-291-0

The Library of Congress has cataloged the original hardcover edition with the following call number: 2010044725

Printed in Heshan City, Guangdong, China.
190P—082011.

CONTENTS

Note to Caregivers:

Throughout this book, many questions are posed to the reader. Some are open-ended and ask what the reader thinks. Discuss these questions with your child and guide him or her in thinking through the possible answers and outcomes. There are also questions posed which have a specific answer. Encourage your child to read through the text to determine the correct answer. Most importantly, encourage answers grounded in reality while also allowing imaginations to soar. Information to help support you as you share the book with your child is provided in the back in the **Additional Notes** section.

Words that are **bolded** are defined in the glossary in the back of the book.

What Shapes the Earth?

The Earth is round, but its surface is far from smooth. Everywhere you look, there are hills and valleys, mountains and gorges, twists and turns. In this book, you will learn about the forces that shape Earth's surface. You will also try to solve a mystery about what created Uluru (pictured above), a mysterious rock formation in Australia.

iScience *PUZZLE*

Why Is This Rock Alone?

Materials
- shoebox
- soil (mostly clay)
- several small objects, such as pebbles and pennies
- thin book
- spray bottle filled with water

This picture shows an ancient rock in Australia called Uluru, or Ayers Rock. There are other rocks like it on Earth. But most of them are part of larger chains of rock. Why is Uluru by itself? Try this simple activity. It may help you answer this question—and other questions throughout the book.

1. Fill a shoebox with an inch of **soil.** This activity works best with a soil that is mostly clay. Clay is made of very small particles. It feels soft and squishy.
2. Press several small objects, such as pebbles and pennies, into the soil. Cover the objects with a very thin layer of soil.
3. Tilt the shoebox slightly by placing a thin book under one end.
4. Use a spray bottle to spray water over the surface of the soil in the higher end of the box. Watch what happens.

5. Continue to spray water and observe the results. Try spraying the water harder or more softly. Use a little water sometimes and a lot of water other times.
6. Do this for several days or even weeks.
Keep track of your observations in a notebook.

7. Answer these questions about your activity.
   - How often did you spray the surface of the soil before you noticed a difference in its appearance?
   - How did the soil change over time?
   - Did the amount of water you sprayed affect the soil?
   - Did the force of the spray affect the soil?
   - How did the pennies and pebbles affect the soil?
   - Use a fan or your breath to create wind. What happens?

DISCOVER ACTIVITY

A Mix of Earth and Water

Materials

- 4 clear plastic cups
- felt-tipped pen for labeling cups
- masking tape
- soil
- sand
- gravel
- modeling clay
- water

Find a spot of bare ground, and dig your fingers in. Imagine someone else doing the same thing in another part of the world. What you feel might be very different from what the other person feels. That's because Earth's surface is covered by many different types of materials, including sand, rocks, clay, and soil.

(clockwise from left): sand, rocks, clay, soil

Soil is made up of tiny bits of rock mixed with pieces of dead plants and animals. Sand is made up of tiny particles of rocks and minerals. Some soils are sandy or contain large amounts of clay. Others are rich with stuff that is or once was alive. This is called organic matter. Potting soil contains a lot of organic matter. Many Earth materials deal with constant bombardment from rain, snow, and other forms of water. What happens when you add water to Earth's materials? Try this activity to find out.

Use the masking tape and felt-tipped pen to label four clear plastic cups A, B, C, and D.

- Put sand in cup A.
- Put **gravel** in cup B.
- Put modeling clay in cup C. This represents the clay on Earth's surface.
- Put potting soil in cup D.

Add an inch of water to each cup. What does the water do in each of these materials?

Rain that falls on streets can cause flooding if it has nowhere to go.

How does seeing how the soil and water interact in the cups help you understand the process that is going on in the shoebox?

Take a look at your shoebox for a minute. There is more room and more dirt in the box than there is in your cups. On Earth there is even more room and more dirt than there is in the box. Activities in a small space such as the shoebox give you a chance to imagine what's happening in the much bigger world.

Consider rain. When water hits Earth's surface, a few things might happen. It can flow away. It can soak into the ground. Or it can collect in pools. What actually happens depends on the type of surface material, the amount of water involved, and the shape of Earth's surface. For example, flooding can occur if there's a lot of water, the land is hilly, the surface is paved, or water doesn't drain well from the surface material.

What is the shape of Earth's surface where you live? What type of material covers the surface? Is most of the surface paved over or exposed?

What Is a Landform?

Even if two places on Earth have the same kind of soil, they might look very different from one another. Hills, plains, lakes, streams, valleys, and other features shape a place. These features are called **landforms.** The landforms in an area make up its landscape.

Mountains are one type of landform.

The Earth didn't always look the way it does now. Its surface changes constantly. Many different forces cause these changes. The landforms you see today may have formed in a few hours or over millions of years. What landforms do you see in this picture? What forces do you think shaped these landforms? How long do you think it took to form each one?

What Causes Landforms to Change?

Changes to the Earth often begin slowly as big rocks break into small, loose pieces called **sediment.** This process is called **weathering.** The changes continue as wind, water, and ice pick up and move pieces of sediment. This is called **erosion.** Eventually, the sediment gets dropped off somewhere else. This is called **deposition.**

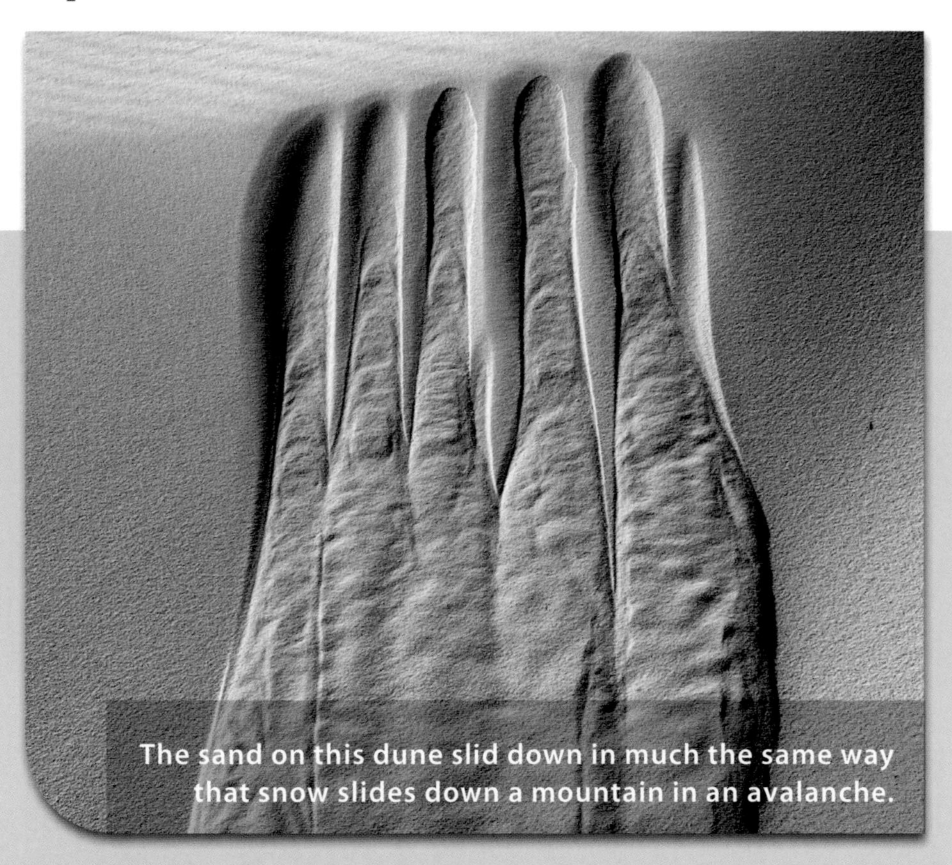

The sand on this dune slid down in much the same way that snow slides down a mountain in an avalanche.

Weathering, erosion, and deposition happen all the time, nonstop. Gravity, a natural force that pulls objects toward Earth's surface, also plays an important role in shaping landforms. How do you think gravity was involved in changes to the landform in this picture? What do you think might make weathering, erosion, and deposition happen more quickly or more slowly?

The sand in this picture is made up of rocks that have been made smaller by weathering.

The Forces of Wind and Water

Weathering and erosion have the power to completely change the way Earth looks. Their main weapons of change are wind and water. Water may take the form of rain or snow. Water may also flow from a source such as a river. Together or alone, wind and water can chisel off particles of rock and carry them away. Over a very long time, these forces make rocks smaller and smoother.

The Mississippi River flows southward 2,320 miles from Minnesota to the Gulf of Mexico.

To see how rivers can shape landscapes, consider this picture of the Mississippi River. Like most rivers, it has bends, or curves. If it were flowing more quickly, it would move basically in a straight path, eroding the land on both sides as it rushed along. A slower-moving river, like this one, swerves around obstacles, such as rocks or bumps in the ground. As it flows around a bend, the water at the outside of the curve moves faster, eroding the land along its sides, called a bank. Sediment then gets dropped by slower water at the inside of the curve. This causes the curve to become even more pronounced. The resulting feature is called a meander. A river that turns is said to be meandering. The Mississippi River flows almost straight in some places, and meanders in others. Where is the water moving quickly and slowly in this picture? How can you tell?

Six of the seven continents have glaciers. Can you guess which continent does not have glaciers?

Ice Sculptors

Water shapes land not just when it's liquid, but also when it's frozen solid. A glacier is a large, slow-moving hunk of ice and snow. Glaciers form where it is very cold and there is a large amount of snow that does not melt. Glaciers can cause weathering, erosion, and deposition when they grow or shrink. Twenty thousand years ago, glaciers covered parts of the United States. When these glaciers shrank, or retreated, they left behind large bowl-like dips in the land, called depressions. The Great Lakes are just one massive example of landforms created by glaciers. What natural forces might cause a glacier to grow or shrink?

Potholes, formed by expanding ice in cracks on roads, are a nuisance to drivers.

Ice Cold Forces

Ice doesn't have to be trapped in a glacier to sculpt landscapes. It can change landforms any time the temperature is below freezing. In cold, wintry places, water can pool on the ground and seep into gravel or cracks in the pavement. When that water freezes to form ice, it expands. This process leaves holes underneath paved roads that get bigger under the weight of cars. Soon the paving collapses, and a pothole results. How could you prevent potholes from forming?

Beyond the Physical

Wind, rain, and ice sculpt Earth's surface in much the same way that you use water and a shovel to make sand sculptures on the beach. It's a physical process, and it's called **physical weathering.** Rock breaks into pieces, but the materials that form the rock don't change. In **chemical weathering,** on the other hand, the materials in the rock do change. For example, some rocks contain a metal called iron.

patina on the Statue of Liberty

rust on metal screw and bolts

When iron is exposed to oxygen, it turns into iron oxide, also known as rust. Heat and water speed up this chemical reaction. When rust forms on the surface of a rock, wind and water can then erode the metal and dump it into the environment.

Any object that contains iron can rust, including cars and wrenches. How can you prevent an object from rusting?

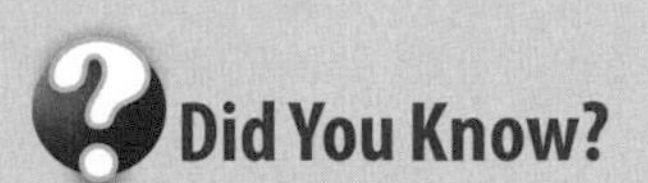

Did You Know?

New York City is home to a famous example of chemical weathering: the Statue of Liberty. The outside of the statue is made of copper. This metal is usually shiny and reddish-brown, like pennies. So why is the Statue of Liberty green? Copper turns green when exposed to oxygen. The greenish coating is called patina. Unlike rust, patina protects the surface of objects from further weathering.

Let's go back and look in your shoebox. Are any of the items showing signs of weathering?

Lava oozes out of a volcano.

Explosive Forces

So far, you've seen a bunch of forces that shape Earth from the outside in. There are also forces that shape Earth from the inside out. For instance, rock that lies deep in Earth is so hot that it melts into a thick liquid. Under the surface, melted or molten rock is called magma. When pressure builds down there, magma may find a path to Earth's surface and can erupt from **volcanoes.** These are mountainous landforms with openings that stretch down to pools of magma. Magma that flows onto Earth's surface is called lava.

When lava cools, Earth takes on a new shape. The **cone** of a volcano is made from cooled lava, dust, and ash. The Hawaiian Islands are an example of landforms made from repeated volcanic eruptions. There, cooling lava has formed layers that have piled up miles above the floor of the Pacific Ocean. How might flowing lava behave like flowing water to shape landforms?

Smoke billows from an active volcano.

As solid as the ground beneath your feet feels, there is actually a lot of action going on down there. Earth's rocky outer shell, or crust, is made of stiff sections called **tectonic plates.** These plates float on magma below. Forces inside Earth cause these plates to move away from each other, rub against each other, or crash into each other.

When plates crash into or move away from each other, magma comes to Earth's surface, forming volcanoes. Other volcanoes grow at places where one plate slides under another plate. Still others grow in what are called hot spots, which are not near the edges of plates. These are thin places in Earth's crust where magma shoots up from deep underground. The Hawaiian Islands have grown over a hot spot.

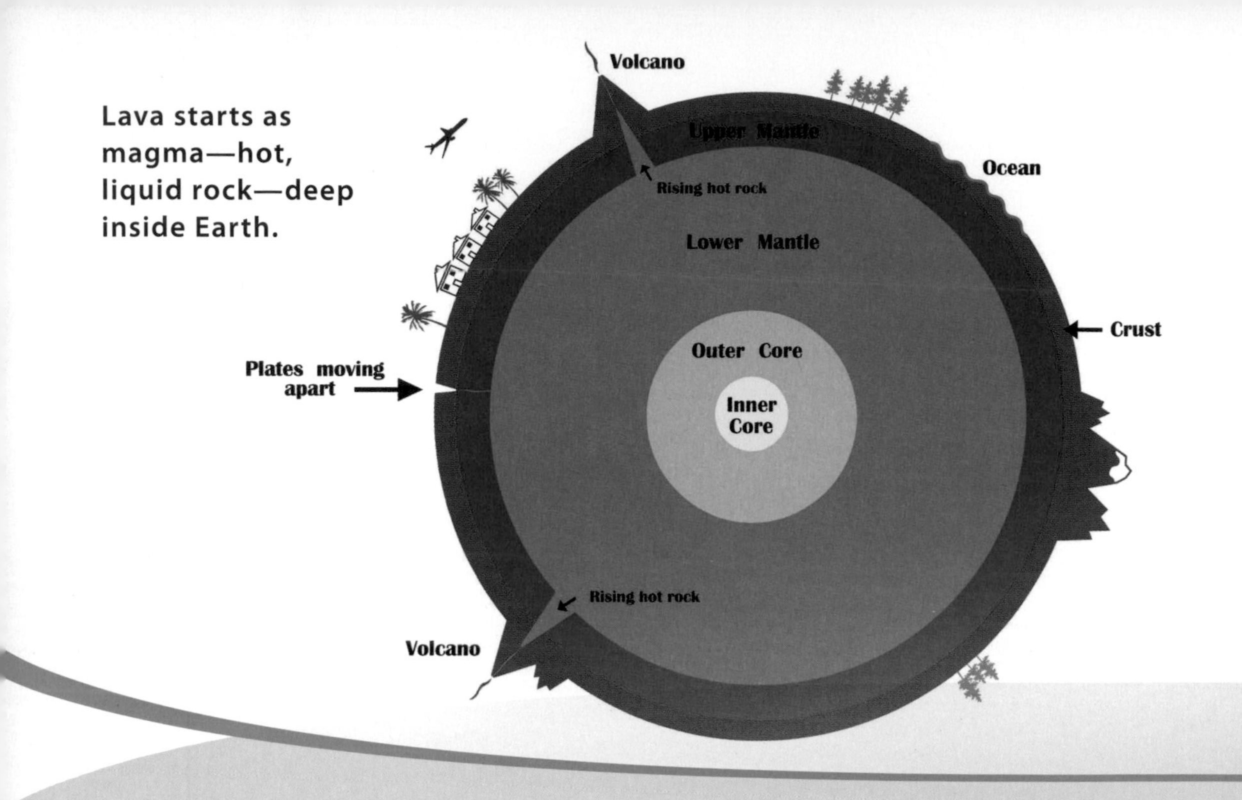

Lava starts as magma—hot, liquid rock—deep inside Earth.

When plates rub against each other, they sometimes get stuck. As the plates finally pull free from each other, the ground can shake. These **earthquakes** often trigger many changes to landforms on Earth's surface.

When plates crash into each other, part of one plate may get crunched and pushed up to form huge wrinkles and folds on Earth's surface. What type of landforms do you think these wrinkles and folds become?

Mount St. Helens before the eruption

CONNECTING TO HISTORY

Mount St. Helens

Mount St. Helens is an active volcano in the state of Washington. Being active means that it can erupt at any time. The volcano's most recent big eruption was in 1980. From March until May that year, more than 10,000 small earthquakes shook the volcano. Steam blasted right through the ice on the top of the mountain. A magma bulge grew along one side, near the top. Scientists knew that something big was about to happen.

On the morning of May 18, a strong earthquake rattled the volcano. Seconds later, the bulge and the top of the volcano erupted. The eruption immediately began to change the shape of the volcano and the surrounding landforms.

Mount St. Helens after the eruption

It was a dramatic eruption. Gas and steam blasted out through the top of the volcano. Magma and rocks created a fierce landslide. All that action destroyed the top north side of the volcano and made a large hole, called a crater. As the landslide traveled down the mountain and into the valley, it dragged trees and soil into streams. Some of these streams were stopped up completely. Others were redirected. Volcanic **debris** and fallen trees also plunged into nearby Spirit Lake, permanently changing its size and depth.

Heat from the volcano melted part of the ice cap at the top of the mountain. Huge volumes of water mixed with crumbling debris to create mudflows. The mudflows raced down the mountain, destroying everything in their path.

Mount St. Helens today

The main 1980 eruption lasted for nine hours. Look back at the pictures of Mount St. Helens before and after the eruption. How is the mountain different? Now look at the picture of the volcano as it is today. How has it changed over time?

Think about your shoebox activity. Suppose you shook it rapidly or sprayed a lot of water into it. How would the contents and layout of the box change? Now try it and see.

How Does Water Change Earth's Surface?

Rumbling, shaking, and exploding forces certainly shape the Earth in dramatic ways. But those dramatic events are rare compared to the day-to-day action of a more simple force: water. Our planet is made mostly of water, and water does its work in many ways.

The force of water carved these rock cliffs.

All of the water that flows on Earth's surface moves downhill. As it flows, water carves channels. A channel is the course a stream of water follows. As time goes on, channels manage and contain the flow of water. The slope, or slant, of a channel (or land surface), may be gradual or steep. Smaller streams of water run together to form larger ones, such as rivers. Where do you think the water in a river ends up?

Think about your shoebox. How could you create a stream in it? Now, try it!

Thornham Creek watershed meets the Atlantic Ocean near Norfolk, Virginia.

Down the Drain

Flowing water has to end up somewhere. Different rivers can drain into different places. Often, a bunch of streams drain into a single area called a **watershed.** A watershed also collects the area's rain or snow as it soaks into the ground and enters streams. All the streams in a watershed run together into a larger river. The river drains into an ocean or another large body of water.

The Mississippi River watershed drains water from nearly half of the continental United States. What river drains your region's watershed? Where does the river end up?

Watershed on the Gulf Coast

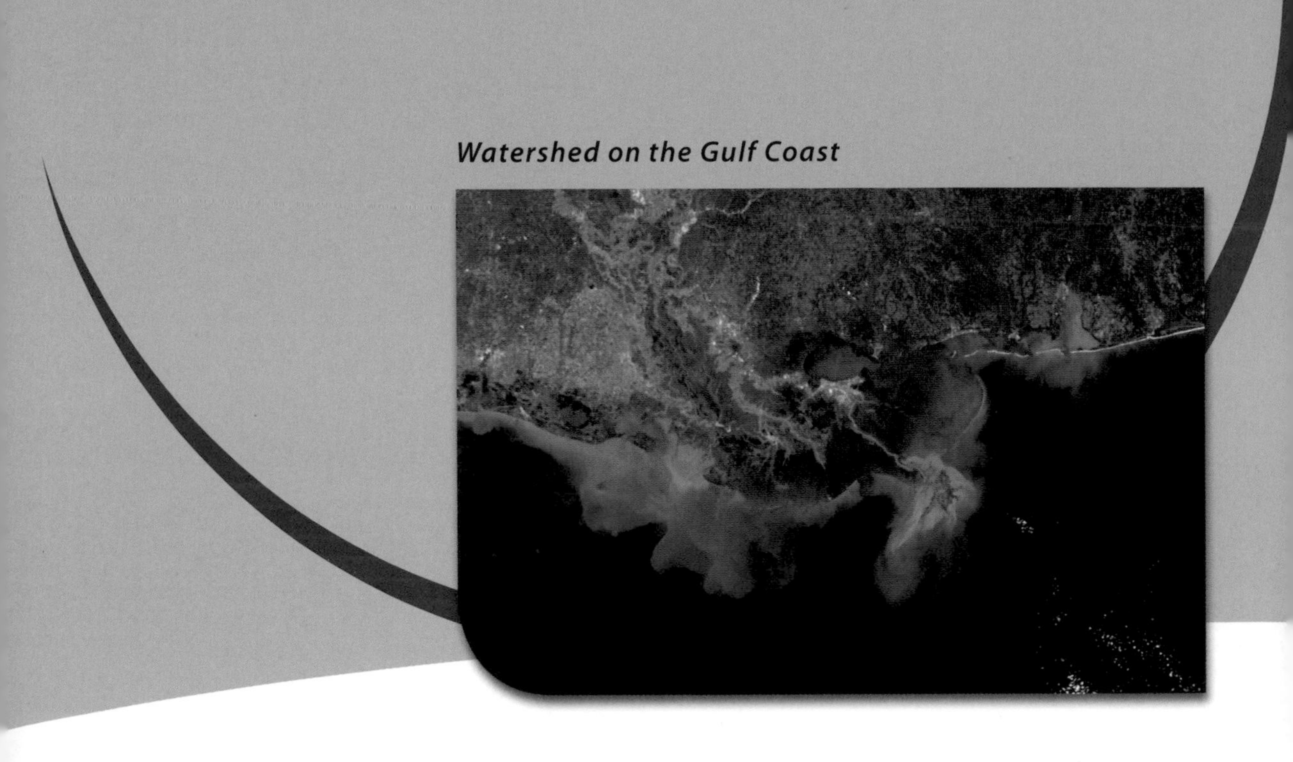

How Does the Water in a Watershed Change Landforms?

Like a whizzing baseball or winning race car, water gains power as it gains speed. Weathering and erosion occur faster. The moving water can carry more sediment. The steeper the slope, the faster the water goes. When the slope is less steep, water tends to move along slowly.

Water also slows down when a river flows into a lake or ocean. As it slows, it drops the sediment it is carrying. This creates a fan-shaped landform in the river mouth called a delta. Can you see the Mississippi River Delta in this picture? What will happen to this delta over time?

Niagara Falls

Over long periods of time, flowing water makes a river channel deeper and wider. This may form a V-shaped valley called a **canyon.** A valley is a low area between higher areas. It took millions of years for the Colorado River to carve the Grand Canyon.

Sometimes, channels lead water over cliffs. Often, these sheer drop-offs form when underground plates shift and tumble. The result is a waterfall. Waterfalls change over time, as the fast-flowing water erodes the rock beneath it. What do you think has happened to the height and location of Niagara Falls over time?

Floods often cause damage to roads, buildings, and vehicles.

How Would a Flood Affect Your Area?

As water builds up, its power grows. Pouring a teaspoon of water over your head wouldn't bother you much, but getting pounded by a bathtub's worth of water would probably really hurt. In the same way, floods can cause a lot of damage to land. A flood is an unusually heavy flow of water over land that is usually dry. Floods have various causes, including hurricanes, heavy rainfall, melting snow, or broken dams.

Some floods, called flash floods, happen within a six-hour period. Heavy rain, snowmelt, or storms can cause them, and they can wash away cars and even homes. They often occur in very dry areas where the water does not soak into the soil.

Often, floods happen when rivers overflow their banks. Floods affect many people and many places every year. An area is never the same after a major flood—for better and for worse. How do you think your area would change after a flood?

Rice fields are carved into step-like terraces so that floodwater can be captured and used to grow crops.

What Is a Floodplain?

Floods have to go somewhere. A **floodplain** is flat land located next to a river. Like a delta, it is made up of sediment that has been dropped off by water. In a floodplain, the sediment comes from rivers that overflowed their banks during floods.

Flooding happens a lot in some floodplains. These floods may make it dangerous for people to live there. However, farmers can use the water from these events to grow certain crops. How else could floods help farmers?

Even calm-looking water can change the shape of a beach over a long time.

How Do Oceans Change Landforms?

Rivers clearly do a lot to shape the land. Oceans, which are much bigger than rivers, change landforms, too. Tides move water onto the shore and away from the shore. This motion both deposits and erodes material on land. When waves crash, they also wear down surface material. As the water retreats to the ocean, it takes stuff back with it.

A beach is sloping land bordering a large body of water that washes over it. Beaches form where sand, stone, and gravel collect. Over time, seawater deposits material on some beaches. Other beaches are being slowly eroded over time.

Water is powerful. Major damage can occur in very little time during a seaside storm.

While those changes are slow and steady, the ocean can also cause quick changes to landforms. Hurricanes or large storms can cause large waves to form. Tsunamis are extra-large, powerful, and dangerous waves that move across the ocean after earthquakes rumble. These events can lead to major changes in the land. And the land itself is not all that can be affected. Buildings, roadways, automobiles, and trees can be damaged. Animals and people can be forced to move and can even suffer injury or death during such severe events.

Ice and glacial activity carved out the depression in this mountain over a long time.

How Does Water Shape Mountains?

In a fight between water and a mountain, which do you think would win? As big as mountains can be, you might be surprised to learn that water is stronger. Rain, snow, and flowing water shape mountains through weathering, erosion, and deposition. Water in the form of ice and glaciers also carves mountainsides.

A mountain usually has steep slopes. Water flows quickly down the sides. As a result, steep mountains erode faster than landforms with more gentle slopes.

Look at the items in your shoebox. Which one seems most like a mountain?

The Smokey Mountains

It usually takes a long time for a mountain on Earth's surface to wear down. As you have learned, mountains form when solid rock thrusts up from Earth's surface. Younger mountains are tall and have ragged tops. As weathering and erosion happen, older mountains become shorter and smoother on top. Which mountains in the pictures on these pages are younger and which are older?

Mapping the Changing Earth

Everyone knows the Earth is round, but most maps are flat. A map is a visual drawing of an area and its features. To make a flat map that shows the shape of the land, mapmakers need information in three dimensions.

A surveyor measures the distance between two points in a neighborhood.

A surveyor is a person who gathers information about the features of landforms, such as shapes and exact locations. Surveyors measure distances by seeing how long it takes to send a signal between two points. Why might it be important to know exact distances on an area of land?

A satellite image shows landforms from space.

green = lowlands
tan = plains
brown = highlands

How Are Maps Made?

Long ago, mapmakers had to create pictures of the Earth based on experience and measurements. Today, mapmakers can see exactly what Earth looks like. They use photographs of Earth taken from high above its surface. Photographs may be taken from airplanes or spacecraft. Satellite images are pictures of Earth's surface taken by satellites in space. Satellites take pictures of a large area and send them to a computer. The computer turns the pictures into a map. Satellite images are useful when landforms can change rapidly, such as during a flood. How else could satellite images be used?

Try this: Put your shoebox on a table and stand above it. Draw a picture of what you see in it. This is your map of the shoebox Earth.

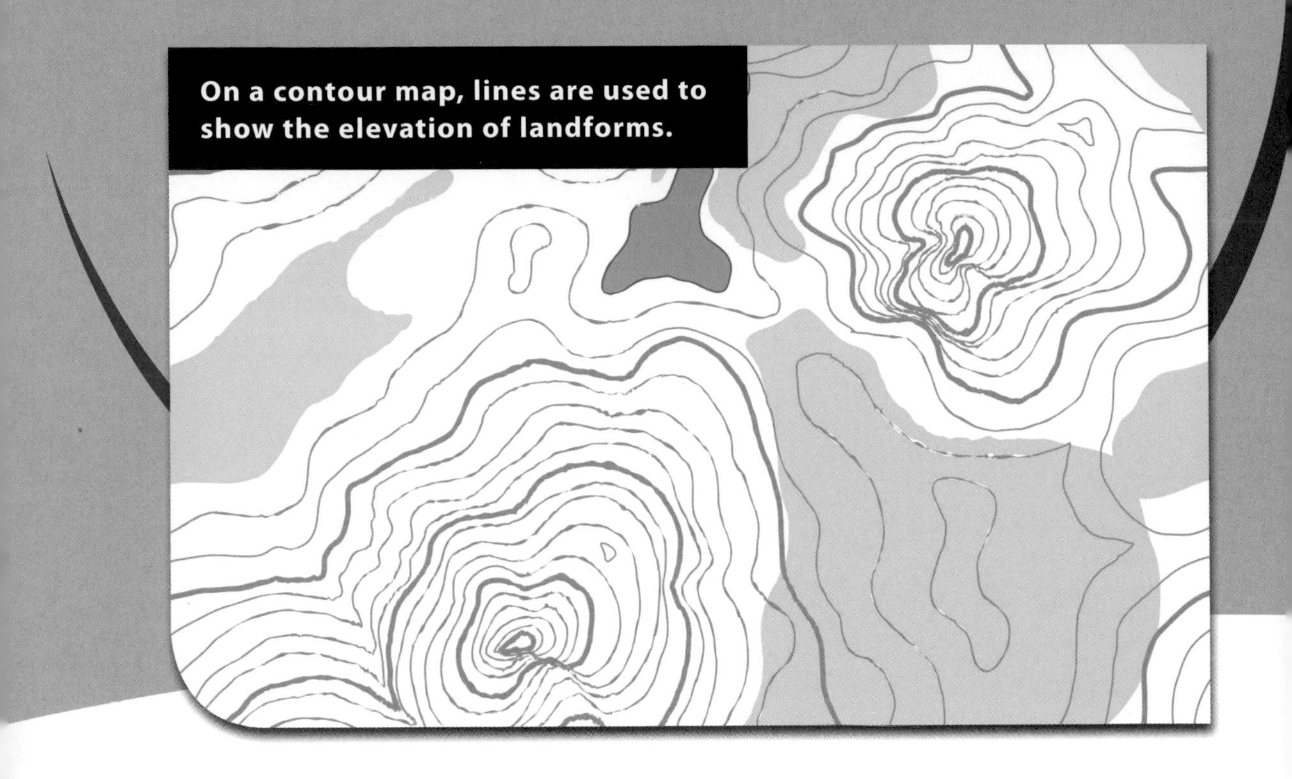

On a contour map, lines are used to show the elevation of landforms.

What Is a Contour Map?

Even a flat map can show hills and valleys. A contour map (or topographic map) is a map that shows the shapes of the landforms in an area. (Bodies of water are shown in blue.) The map uses contour lines to show which parts of the land are at the same height as each other. Numbers on the lines show how high the land is.

On a contour map, high points such as hills appear as a set of irregular circles, one inside the other. The top of a hill or mountain is inside the smallest circle. The circles get bigger (and lower in altitude) as you move out from the middle. The distance between contour lines indicates the slope of the land. When lines are close together, the slope is steep. When lines are far apart, the slope is more gradual. Where are the steepest slopes on this contour map?

Just like the people who built the first cities, most people today live near rivers or oceans. Far fewer people live in distant deserts or on high mountains.

Many people choose to live near rivers, which provide water, food, and transportation.

If you could pick anywhere on Earth to live, where would it be? What would the weather be like? What would the landscape be like? Where would you get things you need, such as food, water, and clothing?

How Do People Change Landforms?

For modern people, life is a constant battle against water, wind, earthquakes, and other natural forces. Large cities need stable land. To keep the land stable, people try to change it. People put up fences to prevent cliff erosion. They flatten land to make it better for farming or building. They pave surfaces and add drainage ditches to make water go where they want it to go.

In places, the Mississippi River flows against a manmade retaining wall that keeps it from eroding the land.

Without this kind of construction, rivers such as the Mississippi would change their paths all the time. The United States has spent billions of dollars building walls, removing dirt from riverbeds, and building dams to try to control the Mississippi River. Do you think the river needs to be controlled?

a farm abandoned during the Dust Bowl era

How Does Farming Affect the Land?

Even farmers, who live off the land, have to fight natural forces to grow their crops. Farmers often need large fields, flat land, and rich soil. A good water supply is also important. Farmers must manage their land so that it will produce crops for many years. To keep the land useful, they sometimes leave a field unplanted for a year to give it a rest. They also plant different crops on a single field in different years. Some farmers use commercial fertilizers, chemicals to fight pests, and different types of watering methods.

In the 1930s, harsh, dry conditions struck the central United States. This era was called the Dust Bowl. Poor farming practices, such as growing one particular crop all the time instead of rotating crops to improve the soil, made the situation even worse. The dry soil was bare and ready to blow away in the wind. The land became useless. The sky was black with dust. People had to leave their homes and search for a new life elsewhere.

Breakwaters are walls that protect shorelines from the power of waves.

How Do People on the Shore Change Landforms?

Living near the sea offers its own challenges. Big waves can damage the land and even buildings. Other dangers include beach erosion and rising water levels. Both can change the shape of shorelines. Sometimes, people build underwater barriers to trap sand and reduce wave action. They may also add sand to a beach to replace sand that washes away. Some people build walls to keep the ocean from reaching their homes and businesses.

Wind on a beach can cause damage, too. The roots of plants growing in sand dunes along a beach help protect the dunes from wind erosion. However, people sometimes clear away plants to use the dunes for recreation and other purposes. What might that do to the dunes and nearby land?

This gardener has just the right soil to grow her vegetables.

SCIENCE AT WORK

Gardeners

Like farmers, gardeners must understand landforms. If the ground is too rocky, water will flow through it too quickly. If the soil is like clay, the water may not drain through it very well. To help water drain and plants grow, a gardener must remove rocks or add other materials.

A gardener also needs to understand the forces that affect landforms. If the land slopes too much, rain can erode the soil. On sloping land, experienced gardeners plant grass or other sturdy plants to anchor the soil. Or they might put wood chips in places where erosion is a problem. On flatter land, using good soil and leveling the surface can prevent erosion and the pooling of rainwater.

Think about your four cups of soil. Which type of soil would be best to use in a garden? Which would be worst? Why?

These students are investigating ways to help protect the land in their community.

How Should People Protect the Land?

No matter where you live, both nature and people affect the land in many ways. To protect the land, groups of people must come together to decide how it should be used. This requires cooperation and research. How is the land in your area changing over time? What are people doing to hurt it? What are people doing to protect it? Now that you know more about the different types of landforms, perhaps you understand why the land must be protected.

Remember that rock standing all by itself in Australia? Based on what you've learned in this book, can you solve the mystery of why it's all alone?

Uluru is made of sandstone. Scientists think that it is all that remains of an ancient chain of mountains. Wind erosion wore down the rest of the mountains. Scientists also think that the reason Uluru is still standing is because the material that makes it up is packed so solidly that wind erosion cannot easily remove it.

Your shoebox activity demonstrates how water erodes materials. You may have noticed that soft materials, such as soil, erode easily. Rock and other hard materials, on the other hand, may be more difficult to weather. As a result, rocky formations may erode more slowly than the surface material around them.

Uluru has withstood wind erosion because of the way it was formed.

BEYOND THE PUZZLE

In this book, you have learned how a variety of forces sculpt and reshape landforms. Take a close look at landforms around you, or at pictures of famous landforms. How do you think they got the shapes they have now? What do you think they may have looked like many years ago? What do you think they will look like in another thousand or million years? Use what you've learned about landforms to think about how your area has changed or will change.

How do you think this landform in the state of Washington was formed? What forces do you think are shaping this area today?

GLOSSARY

canyon: a V-shaped valley.

chemical weathering: the weathering process in which materials in rock are changed.

cone: a volcanic mountain made from cooled lava, dust, and ash.

debris: fragments of materials broken by a destructive force.

deposition: the settling down of eroded sediment.

earthquakes: sudden shaking of the ground.

erosion: the picking up and removal of sediment by wind, water, or ice.

floodplain: an area of low, flat land that is covered by layers of sediment from a river that regularly floods.

gravel: a material made up of pebbles and small bits of rock.

landforms: natural shapes or features of Earth's surface.

physical weathering: the weathering process that breaks rock into pieces without changing the materials that make it up.

sediment: small, loose pieces of Earth's surface material carried away and deposited by wind, water, ice, or gravity.

soil: material made up of rock fragments mixed with dead plant and animal matter.

tectonic plates: the stiff sections of Earth's crust that float on magma below.

volcanoes: landforms, usually mountains, that release lava, gas, and hot ash.

watershed: an area of land drained by a river or a system of rivers.

weathering: the process of breaking down rock on Earth's surface into sediment.

FURTHER READING

Investigating Landforms: Earth and Space Science by Lynn Van Gorp. Teacher Created Materials, 2008.

The Great Mississippi Flood of 1927 (Cornerstones of Freedom, Second Series) by Deborah Kent. Scholastic, 2009 (reprint).

Volcano & Earthquake (DK Eyewitness Book) by Susanna van Rose. DK Children, 2008.

Geology for Kids, **Understanding Landforms, http://www.kidsgeo.com/geology-for-kids/0032-understanding-landforms.php**

ADDITIONAL NOTES

The page references below provide answers to questions asked throughout the book. Questions whose answers will vary are not addressed.

Page 9: In cup A, the sand absorbs the water, but in cup B the water filters down through the gravel and can be seen by looking into the glass. In cup C the water sits on top of the modeling clay before slowly sinking in, and in cup D the potting soil soaks the water up.

Page 12: Gravity pulled material down and away from its source after weathering loosened the material. Severe weather might make weathering, erosion, and deposition happen more quickly. A flood might cause rapid changes in the landscape. Calm weather might cause the changes to happen more slowly.

Page 14: The water moves quickly near the outside edge of the bend, carving it wider; it moves more slowly in the inside of the bend, where sediment has dropped. The water moves more quickly where there are white-tipped waves and more slowly where the river's surface is calm.

Page 15: Falling snow and gravity might cause a glacier to grow; melting snow might cause it to shrink. Caption question: Australia has no glaciers.

Page 16: Seal up cracks in the pavement and don't let water pool on roads.

Page 17: Apply a protective coating, such as paint, to the object to prevent exposure of the iron to oxygen and water. The penny or other items may be showing slight signs of weathering.

Page 18: Just as flowing water can change the shape of the land by adding rivers and lakes, flowing lava changes the shape of the land by adding more surface material to the Earth.

Page 20: They become a mountain chain.

Page 23: After the eruption the top of the mountain has disappeared and left a hole in its place. Over time the sharp features left by the eruption have softened. If you shook the shoebox or sprayed water into it, some of the soil and the objects would move around, and some objects that were buried would now be on the surface.

Page 24: The water in a river ends up in a lake or an ocean. If you tilt the box and pour water in one area at the top, you can create a stream. The stream may leave a channel in the soil, and it may pool as a body of water at the edge of the shoebox.

Page 26: The Mississippi River Delta can be seen as a tan area that starts near the coast and projects out into the Gulf of Mexico. Deposits from the river can be seen in light green under water. Over time the delta will continue to grow offshore.

Page 27: The falls have become taller and moved backward.

Page 29: Floods could deposit sediment that has nutrients good for growing crops.

Page 32: The tallest item might seem most like a mountain.

Page 33: The photo on page 32 shows the younger, jagged mountains. The photo on page 33 shows the older mountains, rounded and softened by weathering and erosion.

Page 34: It is useful for planning travel across or building on that land.

Page 35: They could track weather or other changes across Earth's surface.

Page 36: The steepest slopes are where the contour lines are the closest together. On this map the lines that are closest together are on white areas that appear to be hills or mountains.

Page 38: When the rivers aren't controlled, people's land could end up under a new bend in the river. Boats would have trouble navigating if crews didn't know what to expect. Controlling the river also helps prevent flooding, which can be dangerous and ruin crops and homes.

Page 40: Without plants to help protect the dunes, the dunes might blow away.

Page 41: Potting soil would be the best and clay would be the worst because it would hold too much water.

Page 44: Caption questions: Water probably carved out most of the area long ago. Erosion and weathering from the top are probably shaping the area today.

INDEX